MATH RIDDLES FOR SMART KIDS

400+ Math Riddles and Brain Teasers Your Whole Family Will Love

Cooper the Pooper

© Copyright 2020 Cooper the Pooper - All rights reserved.

The content contained within this book may not be reproduced, duplicated, or transmitted without direct written permission from the author or the publisher.

Under no circumstances will any blame or legal responsibility be held against the publisher, or author, for any damages, reparation, or monetary loss due to the information contained within this book, either directly or indirectly.

Legal Notice:

This book is copyright protected. It is only for personal use. You cannot amend, distribute, sell, use, quote, or paraphrase any part, or the content within this book, without the consent of the author or publisher.

Disclaimer Notice:

Please note the information contained within this document is for educational and entertainment purposes only. All effort has been executed to present accurate, up to date, reliable, complete information. No warranties of any kind are declared or implied. Readers acknowledge that the author is not engaged in the rendering of legal, financial, medical or professional advice. The content within this book has been derived from various sources. Please consult a licensed professional before attempting any techniques outlined in this book.

By reading this document, the reader agrees that under no circumstances is the author responsible for any losses, direct or indirect, that are incurred as a result of the use of the information contained within this document, including, but not limited to, errors, omissions, or inaccuracies.

TABLE OF CONTENTS

Introduction 4

Chapter 1: Scary Riddles 6

Chapter 2: "Which Number?" Riddles 17

Chapter 3: Tricky Riddles 25

Chapter 4: Time Riddles 45

Chapter 5: Age Riddles 55

Chapter 6: Going Shopping Riddles 64

Chapter 7: Easy Riddles 78

Chapter 8: Medium Riddles 91

Chapter 9: Hard Riddles 108

Final Words 126

INTRODUCTION

Well hello there my young mathematician.

I hope I have found you in a good mood on what is going to be a *wonderful* day.

What? How do I know it is going to be a wonderful day?

Well I thought that was obvious.

See, in your hand you hold one of my best pieces of work. You hold a book that is full of my best *math riddles*. Riddles so hard that they will have you scratching your head with frustration. Riddles so good you will be shocked with the answer.

Heck, there might even be a couple of math riddles in there that your parents won't be able to answer (in fact, I know there are).

To be honest, I am very proud of this book.

You might find it hard to believe, but I used to spend all my time playing in the sun, digging through the trash, and running around with the neighborhood kids.

But then something happened.

Something not so good.

Things started to change. Rather than play with *me*, the neighborhood kids started spending all their time glued to the TV.

How boring is that?

As a result, I decided that something needed to change.

So, I got myself organized, put on my thinking cap, and started brainstorming some amazing ways that kids just like you can have *real* fun with your friends and family.

And just like that I had a brilliant idea — I should start writing books.

Books that are full of things you can share with people. Books that get you thinking, laughing, and having fun.

And what better than *math riddles*.

I went ahead and scoured the globe for the best math riddles in the world. Then, once I had found them all, I brought them together into one incredible book (one that my family loves, by the way).

In fact, these riddles are so good they even confused Grandpa — and he is the smartest person I know.

If you are after some great math riddles to share with your friends and family, look no further.

Dive on in, and get ready from some great math riddles.

Just be aware that some of these might have you scratching your head so hard your hair will fall out (I kid, I kid… well, kind of…).

01 A haunted hotel has 500 doors with 500 monsters. The first monster has to go to every door and open it. The second monster has to go to every second door and close it. The third one has to go to every third door, and so on.

- **How many doors are open?**

02 You're in the haunted hotel, and the nearest door to freedom is 100 feet away from you. Each move advances you halfway to the door.

- **How many moves will it take for you to reach the door?**

03 There are 12 zombies in the house. 4 are wearing shoes, and 6 are wearing socks. 3 zombies are wearing both.

- **How many zombies are barefoot?**

04 At our haunted hotel, the currency exchange rate is 1/2 of $5.00=$3.00. If you have 1/3 of $10.00,

- **what is its value?**

05

3 men have decided to stay in our haunted hotel, staying in one room. They pay $30.00 for the room and go up to rest. The bellboy brings their bags along with a $5.00 refund due to a discount. The 3 men gave the bellboy a $2.00 tip and kept the rest of the $5.00. At the end of the weekend, the 3 men decided to tally their expenses but cannot explain why they were coming up with a different figure in their expenses: They each paid $10.00 for the $30.00 room fee. With the room discount, they each got back $1.00, making the room fee $9.00 for each person. They gave the bellboy a $2.00 tip. So their computation goes like this: 3x$9.00+$2.00=$29.00. They cannot account for the missing money.

- **Can you explain what happened?**

06

The King Ghidorah doubles its size every year over 17 years until it reaches its maximum size.

- **How long will it take for the King Ghidorah to reach half its maximum size?**

07

In our haunted hotel there are cockatrices (cockatrice = a two-legged dragon) and rabid dogs. There are a total of 72 feet and 22 heads.

- **How many rabid dogs and cockatrices are in the haunted hotel?**

08 You're trapped in the haunted hotel with 3 magic doors in front of you. 2 of the doors have a monster behind them waiting to pounce on you once you cross the threshold while 1 door will provide you with freedom. You choose door number 2. Door number 1 swings open, and you see it has a monster behind it.

- **Should you change your choice?**

09 Hook has been hired to paint the room numbers on 100 rooms in the haunted hotel, starting from 1.

- **How many times will Hook paint the number 9?**

10 At a vampire convention, 7 vampires meet and shake hands with each of the other vampires once.

- **How many handshakes occurred?**

11 A monster called The Claw has a head that's 6 feet long. Its tail is as long as ½ its body and its head. Its body is ½ of its entire length.

- **How long is The Claw?**

12 There are zombies feasting on unsuspecting guests at our haunted hotel. If 1 zombie feasts on 1 guest, 1 zombie doesn't get a guest. If 2 zombies share each guest, there is 1 extra guest.

- **How many zombies and guests are there?**

13 If 1/4 of the population of monsters in our haunted hotel had 4 legs and the rest of the population had 2, and there are 60 legs total,

- **how many monsters are there?**

14 In Japanese legend, a giant catfish called Namazu lived under the Kashima Shrine and would cause earthquakes every time it moved. If the Kashima Shrine is 300 miles long,

- **how long will the 300-mile-long Namazu take to travel under the shrine if it is moving at 300 miles per minute?**

15 How tall is Gorgo, a prehistoric monster, if his height is 492 feet plus 1/2 his height?

16 Sadako is trying to get out of a well that's 12 feet deep. Every time she is able to climb up 3 feet, she slides back down 2 feet. Climbing up and sliding down takes 1 minute.

- **How many minutes does she take to get out of the well?**

17 If a sasquatch can double its size daily, and, when placed in a cage, it can fully fill the cage in 10 days,

- **how many days will it take for the sasquatch to fill ½ and ¼ of the cage?**

18 If 3 monster hunters can catch 3 monsters in 3 minutes,

- **how long will it take 100 monster hunters to catch 100 monsters?**

19 A baby monster weighs 294 kilos divided by ⅙ of its weight.

- **How much does it weigh?**

20 A mother vampire has a 24-ounce container full of blood that she will be giving to her 3 children, Ariana, Nathan, and Tammy. She only has 5, 11, and 13-ounce bottles that her children can use to drink from. Her children are finicky and refuse to drink from the same bottle.

- How can she divide the blood equally so each child gets the same amount but not have any 2 kids drink from the same bottle?

21 A boobrie is a shape-shifting creature. It normally takes on the form of a giant waterbird. It has a head that is 9 feet long. Its tail is $1/2$ the size of its body, plus the size of its head. Its body is the size of the tail plus the head.

- How long is the boobrie?

22 Out of 100 monsters in the haunted hotel:
85 had fangs
75 had hooves
60 had fur
90 had talons

- How many monsters must have had all 4?

23

There are 100 pairs of monsters in the haunted hotel. 2 pairs of baby monsters are born for each and every monster. 23 of the baby monsters did not survive.

- **How many monsters in total are in the haunted hotel?**

24

In a monster club, there are 600 members. 5% of the members have 1 eyeball. Of the 95% remaining, ½ have 2 eyeballs, while the other ½ do not have eyes.

- **How many total eyeballs are there in this club?**

25

There is a dungeon in our haunted hotel. In the dungeon, there is a chain nailed to the wall. This chain is 10 meters long, and its center dips down 5 meters from where each side of the chain is nailed to the wall.

- **How far are the ends of the chain from each other?**

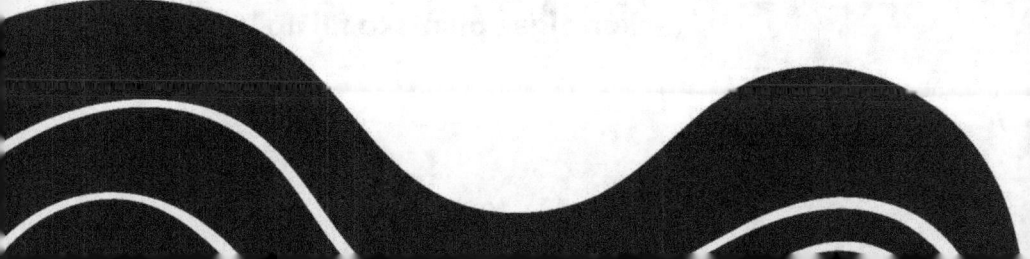

26

100 rooms, numbered 1 to 100, in our haunted hotel have been booked by guests. The head monster goes to each of the 100 rooms, checks if it is ready for the guest the next day, and closes the door. 100 poltergeists decided to play a trick on the head monster. The first poltergeist opens every door to the 100 rooms. The next poltergeist closes every other room starting at the second room. Poltergeist 3 goes to every third door starting with the third door and opens it if it's closed and closes the door if it's open. Poltergeist 4 opens every fourth door if its closed and closes the door if it's open. This continues up to poltergeist 100, who goes to Room 100. When the head monster checks in on the rooms again the next morning,

- which rooms have their doors open?

27

Alien algae landed in a nearby lake and has been doubling its size each hour. It will take 18 hours for 1 plant to fill up ½ the lake.

- How long will it take for 2 of these alien algae plants to fill up ½ the lake?

ANSWERS

1. 22
2. You will never reach the door if you always move just half the distance.
3. 5
4. $4.00
5. The 3 men spent $27.00.
6. 16
7. 8 cockatrices and 14 rabid dogs
8. Yes, you have a 66% chance of success by changing your door choice.
9. 20
10. 21
11. 48
12. 4 zombies and 3 guests
13. 24
14. 2 minutes
15. 984
16. 10
17. 8 days to fill ¼ of the cage and 9 days to fill ½ of the cage
18. 3
19. 42
20. Fill the 11 and 5-ounce bottles, leaving 8 ounces in the 24-ounce bottle. Give this to Ariana. Pour the blood from the 11-ounce bottle into the 13-ounce bottle, and top it up with 2 ounces of blood from the 5-ounce bottle. Pour the remaining 3 ounces of blood in the 11-ounce bottle. Pour 5 ounces of blood from the 13-ounce bottle into the 5-ounce bottle, leaving 8 ounces of blood in the 13-ounce bottle. Give this to Nathan. Pour the blood from

the 5-ounce bottle into the
11-ounce bottle and give this
to Tammy.
21. 72
22. 10
23. 977
24. 600
25. Both ends are attached to the same nail, so the distance is 0.
26. Rooms 1, 4, 9, 16, 25, 36, 49, 64, 81, and 100
27. 17

CHAPTER 2
"WHICH NUMBER?" RIDDLES

01 I am a number with a pair of friends. You'll find me once more with a ¼ of a dozen.
- **Which number am I?**

02 Add me to myself. Multiply that sum by 4. Divide me by 8, and I will appear once more.
- **Which number am I?**

03 I am the number you get when you multiply all the numbers on the telephone keypad.
- **Which number am I?**

04 Which 3 positive numbers have the same results when you add or multiply them together?

05 I occur the most frequently between 1 to 1,000, including 1,000.
- **Which number am I?**

06 I occur the least frequently between 1 to 1,000, including 1,000.
- **Which number am I?**

07 We have a two-digit answer when added together and a single-digit answer when multiplied.
- **Which whole numbers between 0-10 are we?**

08 I am a three-digit number. My ones digit is 5 less than my tens digit. My tens digit is 8 more than my hundreds digit.
- **Which number am I?**

09 The square of the first number and the square of the second number equals 8. The first number plus the product of the second and first number equals 6.
- **Which 2 numbers are we?**

10 My last digit is 4 times the first digit. The first digit is 1/6 the second digit. The second digit is 3 less than the third digit.
- **Which four-digit number am I?**

11 Which 2 numbers have a difference of 24 between their squares, and their product added to the sum of their squares is 109?

12 If you add me to my square, and you add the digits of that summation together, you arrive back at me (the original number).

- **Which number am I?**

13 I have 3 digits. Divide me with 36 and the product equals the sum of my 3 digits. 5 times the sum of the middle and right digit equals 7 times the left digit plus 9. The sum of the left and right digits equals the middle digit times 8 minus 9.

- **Which number am I?**

14 We are 4 numbers. Our sum equals 45. If you add 2 to the first number, subtract 2 from the second number, multiply the third number by 2, and divide the fourth number by 2, the results will all be the same number.

- **Which 4 numbers are we?**

15 Which four-digit number am I if the last digit is 5 times the first digit, and if you multiply the last digit by 3 you'll get the second and third digit?

16 Which number am I so that when I add myself to either 100 or 164, the sums are perfect squares?

17 123,456,789,8765,4321 is what number squared?

18 Which set of 3 whole numbers' sum equals their product when multiplied?

19 I am the smallest whole number that equals 7 times the sum of its digits.
- **Which number am I?**

20 Which number is the smallest one that increases by 12 when it is turned upside down and flipped?

21 In the equation below, each letter represents a different number, none of which is 0.

- **Which numbers do each letter represent?**
ABCDEx4=EDCBA

22 Which number multiplied by 4 gives the same number but in reverse?

23 In this sequence, which number does the question mark represent?

- 7, 14, 42, 168, ?

24 Add me to myself, and then multiply the sum by 4. Divide that number by 8, and you will get me again.

- **Which number am I?**

25 If you multiply me with any number, you will always get the same answer.

- **Which number am I?**

26 Which three-digit number has its ones digit 5 less than its tens digit and its tens digit is 8 more than its hundreds digit?

27 Which number do you get when you divide 30 by ½ and add 10?

28 I am the smallest possible three-digit palindrome that you can divide by 18.
- **Which number am I?**

29 Which number goes at the end of this sequence?
- 1, 11, 21, 1211, 111221, 312211, …

30 Subtract this number with 100 and you get 0. This number is not higher than 100, but not less than 90. This number is not an odd number.
- **Which number is it?**

ANSWERS

1. 3
2. Any number
3. 0
4. 1, 2, 3
5. 1
6. 0
7. 9 and 1
8. 194
9. 2 and 2
10. 1,694
11. 5 and 7
12. There are 3 possible answers: 0, 3, and 9.
13. 324
14. 8, 12, 5, 20
15. 1155
16. 125
17. 111,111,111
18. 1, 2, 3
19. 21
20. 86
21. A=2
    B=1
    C=9
    D=7
    E=8
    4A has to be less than 10, and A is an even number, so A=2, then making E=8. Ex4 has to end in 2, so B=1. D≥4 and D8x4 ends in 12, so D=7. C=9 then.
22. 2178
23. 840
24. Any number
25. 0
26. 194
27. 70
28. 252
29. 13112221
30. 100

01

If:
19,235=1
123=0
1,001=1
4,235=0
8,890=6
656=2
5,390=2

- Then what does 123,456,789 equal?

02 How is it possible for someone to be 20 years old in 1980 and 15 years old in 1985?

03 How much is ½ of 2 plus 2?

04 What are the chances of a coin landing tails up if it has landed heads up the previous 10 tosses?

05 How can you make the number 7 even without division, subtraction, multiplication, or addition?

06 A rabbit can jump forever but gets tired with each jump and can only jump ½ as far as her previous jump. At her first jump, she is able to jump ½ a foot. Her next jump, she is able to jump ¼ of a foot.

- **How many jumps will it take to jump 1 foot?**

07 How can you take 2 from 5 and get 4?

08 How many times can you take away 5 from 25?

09 When Bryan was 6 years old, he hammered a nail at the base of his favorite tree. If the tree grows 5 centimeters each year, and Bryan returns after 73 years,

- **how high up would the nail be?**

10 A clothing store in Oklahoma charges $30.00 for a blouse, $20.00 for a vest, $15.00 for a tie, and $25.00 for socks.

- **How much will underwear cost?**

11 What did the triangle say to the argumentative circle?

12 How do you make this equation true?
81 x 9 = 801

13 Make this equation true by drawing just 1 straight line: 5+5+5=550

14 If you look at a 38° angle measure under a microscope that is magnified at 10 times,
- how much will the angle measure then?

15 Why is it easier to count cows compared to sheep?

16 What happened to the vegetation in the math classroom?

17 What are 2 math friends called?

18 What is the handiest tool in a math teacher's toolbox?

19 I am a place in one of the most famous cities in the world. I am also an answer to this math problem:
- **what do you multiply by 6 to get a cube?**

20 What are 5 and 5, if 2 is company and 3 is a crowd?

21 Why are 1993 fifty dollar bills worth less than 1999 fifty-dollar bills?

22 A truck crosses a 10-mile-long bridge that can only hold 14,000 pounds, the exact weight of the truck. Halfway through, a bird lands on the truck.
- **Will the bridge collapse?**

23 A ping pong ball is thrown off the roof of a 90-foot building. Each time it bounces, it goes back up halfway.
- **This ball will bounce how many times before it stops?**

24 David is walking down the street one night. As he passes a street light, he sees his shadow get longer. Assuming he is walking at a constant pace, will his shadow move faster, slower, or the same when his shadow is longer as when it is shorter?

25 Which bag is worth more: a bag with a pound of $10 silver pieces or a bag with half a pound of $20 silver pieces?

26 3 sparrows fly outward from their nest. When will all 3 sparrows be on the same plane in space?

27 How many gila monsters are in a small square room if, in each corner, there is a gila monster sitting, opposite each sit 3 gila monsters, and 1 is sitting near each one's tail?

28 For a range of numbers between 1 to 19, what are the next 5 numbers in this sequence?

- 8, 18, 11, 15, 5, 4, 14, 9, 19, 1, 7, 17, 6, 16,?,?,?,?,?

29 In this series, what is the answer to the question mark?

1	3	5
2	4	?

30 How can you make 9+5=2?

31 If there are 4 biscuits and you take away 1,
- **how many do you have?**

32 What can you put between 5 and 6 so that the number is greater than 5 but still less than 6?

33 You have 2 coins whose total value is $0.30. One of them is not a 5-cent coin.
- **What are the values of the coins?**

34 Ari bought a rooster so she can have eggs for breakfast. She expects to get 3 eggs every day. After 3 weeks,
- **how many eggs will she have had?**

35 If it took 12 people 5 hours to build a barn,
- how long will it take 6 people to build the same barn?

36 Selena has 2 books wherein 1 is upside down and the other is rotated so the top is facing her. In this case,
- what is the total sum of the first pages of these books?

37 Nathan has a pound of steel and a pound of cotton.
- Which one weighs more?

38 What is the answer to the question mark?
12x12=9
23x23=16
34x34=?

39 There is an empty container that is 1 meter in diameter.
- How many coconuts can you put in this empty container?

40

Complete the sequence

1=3, 2=3, 3=5, 4=4, 5=4, 6=3, 7=5, 8=5, 9=4, 10=3, 11=?, 12=?

41

If you had 5 coconuts and 2 durian in one hand and 2 coconuts and 4 durian in another hand,

- **what would you have?**

42

A yoyo has a blue side and a black side. There are 100 yoyos in a box. The box breaks, and the yoyos scatter everywhere. 90 of these yoyos fell with the blue side facing up, and the other 10 fell with the black side facing up. Your task is to sort these yoyos into 2 piles with each pile having the same count of yoyos with the black side facing up.

- **How do you do this?**

43

Charlie was on his way to Atlanta. On the way, he met Sally, who had 7 children with her. Each of these children had 7 dogs with them, and each dog had 7 puppies with her.

- **How many legs will be going to Atlanta?**

44 If 1+9+8=1, then what is 2+8+9?

45 What never occurs in a day but once in a year and twice in a week?

46 There are 4 days in the week that start with the letter T.
- **Name them.**

47 How did the basketball fan know before the game that the score will be 0-0?

48 Charlie died of old age on his 25th birthday.
- **How did this happen?**

49 Katie has 17 cheetahs, and all but 9 die.
- **How many cheetahs are left?**

50 What does 816,982 equal if the following are true:
66=2
999=3
8=2
0=1
8,123=2
98=3
88=4

51 Kesha was born in 1973 and passed away in 1986 at the age of 43.
- **How did this happen?**

52 Taylor got a flat tire while driving down the highway. She took off the flat, but before she could put on the spare, all the wheel nuts got blown away down a cliff by a strong wind.
- **How can Taylor attach the spare tire now?**

53 Which animal has a cent?

54 What do a dollar and the moon have in common?

55 Selena is able to hold her breath under water for 15 minutes. A 10-year old approached her and made a bet for $1,000.00 that she could stay under water for 30 minutes. The 10-year old did and walked away with the $1,000.00.

- How did she do it?

56 An old billionaire had no one to leave her assets to, so she told people in her town that she would leave her money to the person who can pet her dog half as many times as the days she had left to live. Since no one had any idea when she was going to die, no one attempted the challenge, except for Selena who ended up getting all the billionaire's money when she passed away.

- How did Selena do it?

57 What did the number 2 say to number 1?

58 I work day in, day out, 24 hours a day, every hour, every second, every minute.

- What am I?

59 How can 2 people born simultaneously have different birthdays?

60 What is the smallest amount of time?

61 What never comes in a day but twice in the afternoon and once in a second?

62 What has 4 feet and is flat but cannot walk?

63 What do you get when you add 5 apples and 2 blackberries?

64 Which is the largest: rectangle, triangle, square, or circle?

65 How do you fit 4 apples into a container that is 7x4x2 inches?

66 Why shouldn't anyone mention the number 288 to anyone?

67 Bar another number, what can be added to 3 to turn it into an even number?

68 What did the plus sign say to the minus sign when they saw each other on Facebook?

69 This never occurs in May, twice in November, and once in June.
- **What is it?**

70 What five-letter word becomes larger when you add one letter to it?

71 If hexagons have 6 sides, pentagons have 5 sides, decagons have 10 sides, and octagons have 8 sides, a family of 3 have how many sides on their dining table?

72 What is notable about the number: 8,549,176,320?

73 Where can you buy a three-foot long ruler?

74 Which coin doubles its value when you deduct half?

75 What has no toes and 3 feet?

76 Why are $100 bills like diapers?

77 Why is the longest human nose recorded only 11 inches?

78 Why don't they serve alcohol at math parties?

79 Why didn't the quarter roll down the mountain with the nickel?

80 What is a proof?

81 What did one math book say to another?

82 What's wrong if I said I had a girlfriend that's the square root of -100? She's a perfect 10.

83 You're on a bus with 10 passengers. At the first stop, 7 get off and 2 passengers get on. At the next stop, 2 get off and 8 get on. At the next stop, 4 more get off and 2 get on. At the last stop, no one gets on or off.

• **How many people are on the bus?**

84 A tomato vine has grown to 3 meters long. A tomato grows every 5 inches except for the bottom foot of the vine.

• **How many vegetables grow off the vine?**

85 What is bigger than anything you can think of? You can find it behind stars, or in a third, a sixth, or a seventh. It takes a computer, something round, or this to make pie.

86 1+9+8=1
2+8+9=?

87 Which is correct, 5 and 9 is 13 or 5 plus 9 is 13?

88 If 3 containers have the same amounts of pennies, dimes, and quarters,
- which container will have the highest value?

89 Greg has piles of leaves. There were 3 in the front yard, the backyard had 4, and at the side of the house there were 5 piles.
- How many piles of leaves will he have if he combined all of them in the front yard?

90 Vanessa is up in a mango tree using a pot.
- What do you call this pot?

91 Expand $(a+b)^2$

ANSWERS

1. 4
2. The person was born in 2000 B.C. (Before Christ).
3. 3
4. 50/50
5. Drop the "s."
6. Never
7. Remove the letters "F" and "E" so you're left with IV.
8. Once
9. The nail would still be at the base of the tree since a tree grows at the top.
10. $45.00
11. You don't have a point.
12. Turn it upside down.
13. There are 2 possible answers: You can draw a line through the equal sign to make it "not equal to" or you can draw a line on the plus sign to make it 4 so the equation reads 545+5=550.
14. 38°
15. Because you can use a cowculator.
16. It grew square roots.
17. Algebros
18. Multi-pliers
19. Times Square
20. Nine
21. Because there are less of them.
22. No. The truck would've lost weight from the consumption of gas from the 5 miles it has traveled over the bridge.
23. The ping pong ball will bounce infinitely.
24. The movement of the shadow will be at a constant speed, and the length of the shadow has no bearing on how fast it will move.
25. The bag with the pound of $10 silver pieces.

26. Always, as there is always a plane in space that contains any 3 points.
27. Four gila monsters near the tail of one in an adjacent corner.
28. 8, 18, 11, 15, 5, 4, 14, 9, 19, 1, 7, 17, 6, 16, 10, 13, 3, 12, 2
29. R
30. If it is 9AM, add 5 hours and you get 2PM.
31. One
32. A decimal point
33. $0.25 and $0.05
34. None
35. It will take no time since the barn is already built.
36. 2
37. Both weigh the same.
38. 13
39. One coconut
40. 6
41. Extremely large hands
42. It was never indicated that the piles had to be of the same size, so you can make a pile of the 90 yoyos in one and 10 in the other. If you flip all 10 yoyos in the second pile, you will have the same count of yoyos with the black side facing up.
43. Two — only Charlie is going to Atlanta.
44. 10
45. The letter E
46. Tuesday, Thursday, Tomorrow, and Today
47. In any basketball game, the score will always be 0-0 before it starts.
48. Charlie's birthday is on February 29.
49. 9
50. 6
51. Kesha was born in room 1973 and died in room 1986 of the same hospital.
52. Take off one wheel nut from the other 3 tires so each tire has 3 nuts.
53. A skunk has a scent.
54. Four quarters
55. The 10-year old held a glass

of water over her head for 30 minutes.
56. Selena went to pet the dog every other day.
57. Are you single?
58. A clock
59. This will happen when one is born in one time zone different from the other.
60. Minute
61. The letter O
62. Tape measure
63. A house full of gadgets
64. Rectangle because it has the most letters
65. Put 4 iPhones in the box.
66. It is too gross.
67. E can turn it into the number 8 if you connect them.
68. Add me.
69. The letter E
70. Large
71. It depends on the number of side dishes being served.
72. The digits 0 to 9 are listed in alphabetical order.
73. At a yard sale
74. A half-dollar
75. A yardstick
76. They need change.
77. It would be a foot otherwise.
78. Because you should never drink and derive.
79. Because it had more cents.
80. ½ percent of alcohol
81. You're not the only one with problems.
82. Yes, but purely imaginary
83. 10, including the driver
84. None; tomatoes are fruit.
85. Infinity
86. 10
87. Neither
88. The container with the quarters
89. One, since they're combined
90. High-pot-in-use
91. $(a+b)^2$

01 What time is it if 2 hours later, it takes ½ as much time until it's noon as it would 1 hour later?

02 A clock chimes 5 times in 4 seconds.
- How many times will it chime in 10 seconds?

03 A tree doubled in height each year until it reached its maximum height over the course of 10 years.
- How many years did it take for the tree to reach ½ its maximum height?

04 A 300 ft. train is traveling 300 ft. per minute through a 300 ft. long tunnel.
- How long will it take the train to travel through the tunnel?

05 If it took 8 men 10 hours to build a wall,
- how long would it take 4 men to build it?

06 I possess 2 wires. Both of them have an inconstant thickness, but both of them burn completely in 60 minutes. The problem is that I want to measure 45 minutes while using these 2 wires.
- **How can I measure if cutting the wire in half is not possible due to non-homogeneous construction?**

07 I drive at an average speed of 30 miles per hour to the railroad station each morning and just catch my train. On a particular morning, there was a lot of traffic, and at the halfway point I found I had averaged only 15 miles per hour.
- **How fast must I drive for the rest of the way to catch my train?**

08 How many hours are there in 90 minutes?

09 If 3 salesmen can sell 3 cars in 7 minutes,
- **how many cars can 6 salesmen sell in 70 minutes?**

10 A clock seen through a mirror shows 8 o'clock.
- **What is the correct time?**

11 What time would it be now if 2 hours ago it was as long before one o'clock in the morning as it was after one o'clock in the afternoon?

12 Alicia agreed to work on a job which takes 30 hours to do. The job will pay $8 an hour, on the condition that she will pay her employer $10 for each hour she loafs around. At the end of the 30 hours, Alicia wasn't paid anything, but didn't owe anything either.

- **How many hours did she work, and how many hours did she loaf around?**

13 If it takes the grandfather clock 6 seconds to strike at 4AM,
- **how long will it take to strike at noon?**

14 How many times does the minute hand of a clock lap the short hand between midnight one day and midnight the next day? The starting point of midnight does not count as a lap.

15 When can you add 2 to 11 and get 1 as the answer?

16 Justin discovered that his watch gets affected by sharp changes in temperature. He traveled to the Snake River Plain, which has great diurnal temperature variations. In the daytime, the temperature went as high as 100°F, and at night it went down to -50°F. This made Justin's watch go half a minute faster at night, but it lost 20 seconds at dawn, in effect making it only 10 seconds fast. On the morning of May 1st, his watch showed the correct time.

- **By what date was his watch 5 minutes faster?**

17 The cuckoo clock struck 6 times at 6AM. The time between the first cuckoo and last cuckoo was 30 seconds.

- **How long will the cuckoo clock take to strike at noon?**

18 You planted sunflower seeds in your back garden. Every day, the number of flowers doubles. If it takes 52 days for the flowers to fill the garden,

- **how many days would it take for them to fill half the garden?**

19 How many years are there in a decade?

20 A clock strikes once at 1 o'clock, twice at 2 o'clock, 3 times at 3 o'clock and so on.
- **How many times will it strike over the period of 24 hours?**

21 3 bells toll at the intervals of 10, 15, and 24 minutes. All 3 begin to toll together at 8am.
- **At what time will they toll together again?**

22 A clock runs 4 minutes slow each hour. Its time was set right 3 ½ hours ago. Right now, it is noon.
- **How long will it take the clock to show noon?**

23 How can the face of a clock be divided by 2 straight lines so that the sums of the numbers in each division are equal?

24 In your bookshelf you have 5 favorite books. If you decide to arrange these 5 books in every possible combination and move just 1 book in every half a minute,
- **how much time will it take you to arrange them?**

25 What time is it when the clock strikes 13?

26 Drake has 2 friends living on opposite sides of town. He makes time to see them every Saturday. He goes to the nearest bus station near his house at a random time every Sunday and hops on the next bus that arrives. If the bus goes east, he'll visit Zoe. If the bus goes west, he'll visit Joe. Both east and westbound buses run every 20 minutes. After a few months, Joe started complaining that he only sees Drake once every 5 Saturdays.

- **How is this possible if Drake is going to the bus station at random times each Saturday?**

27 How many months are there in a century?

28 A clock seen through a mirror shows quarter past 3.
- **What is the correct time shown by the clock?**

29 I add 5 to 9 and get 2. The answer is correct,
- **but how?**

30 A clock shows the time as 12:20.
- What is the angle the hour hand makes with the minutes hand?

31 What did 12 o'clock say to 1 o'clock when they kissed?

32 I am shorter than my friends and change a little every 4 years.
- What am I?

33 Tom was not at school last Saturday. He was first absent for 4 days before that. Today is Monday, the 31st of October.
- When was Tom first absent? Give the day and date.

34 How many minutes will it take to fill a bathtub if you accidentally didn't plug it and have both hot and cold taps full on? The water empties in 4 minutes, the cold tap takes 2 minutes to fill the bath, and the hot tap takes 6 minutes.

ANSWERS

1. 9
2. 11 times. It chimes at zero and then once every second for 10 seconds.
3. 9 years
4. 2 minutes. It takes the front of the train 1 minute, and the rest of the train will take 2 minutes to clear the tunnel.
5. No time at all; it is already built.
6. It is not hard as it seems. I will burn 1 wire on both ends and the other wire at 1 end only. The first wire will burn completely in 30 minutes, and at that very moment, I will burn the other end of the second wire, and it will burn in 15 minutes. Both of the wires will be burned in 30 + 15 = 45 minutes.
7. The train is just about to leave the station, and there is no way I will be able to catch it this time.
8. 1.5 hours
9. 60 cars
10. 4:00
11. 9
12. Alicia worked for 16 $\frac{2}{3}$ hours and did nothing for 13 $\frac{1}{3}$ hours.
13. 22 seconds
14. 21
15. When you add 2 to 11 o'clock, you get 1 o'clock.
16. May 28
17. 66 seconds
18. It would take 51 days. If the number of flowers doubles every day, half the garden would be full the day before, on the 51st day.

19. 10 years
20. 156 times
21. 10AM
22. 15 minutes
23. The 2 lines cannot cross, and there will be 3 divisions on the face of the clock. Each part will have the sum of 26. Group the pairs that add up to 13 (e.g. 8+5 and 7+6; 9+4 and 10+3; 11+2 and 12+1)
24. 1 hour
25. Time to get the clock fixed.
26. While both buses arrive at the bus station near Drake's house every 20 minutes, the westbound bus comes 4 minutes after the eastbound bus, decreasing Drake's chances of getting on the westbound bus.
27. 1200 months
28. 8:45
29. When it is 9AM, add 5 hours and you will get 2PM.
30. 110
31. Your first time?
32. February
33. Tuesday, Oct. 25th
34. 2 minutes and 24 seconds.

01 In 2 years, Taylor will be twice as old as 5 years ago.
- **How old is Taylor now?**

02 If Angel was half as old as Tippy 4 years ago, and 4 years from now, Tippy will be 1 1/3 of Angel's age.
- **How old is Tippy?**

03 When Miguel was 6 years old, his little sister, Leila, was 1/2 his age. If Miguel is 40 years old today,
- **how old is Leila?**

04 Jack said, "When Gwen is twice as old as Dean, then I shall be just 17. But Gwen was 23 when Dean was twice as old as me."
- **How old was Jack when Dean was 10?**

05 Charlotte is 13 years old. Her father, Montague, is 40 years old.
- **How many years ago was Charlotte's father 4 times as old as Charlotte?**

06

When I asked her how old she was, she smiled and said cryptically, "The day before yesterday I was 22, but next year I'll be 25."

- **What is her birthday, and when was the date of our conversation?**

07

A boy and his big sister are sitting around the kitchen table chatting. The boy says, "If I take 2 years away from my age and give them to you, you'd be twice my age." The sister adds, "Why don't you just give me 1 more and then I'll be 3 times you age."

- **How old are the siblings?**

08

When Ashley was 15, her mother was 37. Now, her mother is twice her age.

- **How old is Ashley?**

09

John is twice his brother's age and is ½ the age of his father. In 50 years, his brother will be half the age of his father.

- **How old is John now?**

10 How many years ago was Dell's mom 3 times Dell's age if her mom is 80 years old and Dell is 54?

11 If Akon is 4 times as old as his daughter,
- **how old are they both now if, in 20 years, Akon will be twice as old as her?**

12 When Andy was 8, his mother was 31. Now his mother is twice as old as Andy.
- **How old is Andy now?**

13 How old are Beanie and Maya if the sum of their ages is 49? Beanie is twice as old as Maya was when Beanie was as old as Maya is now.

14 I asked Nate how old he was, and he said, "Next year I will be 25, but the day before yesterday, I was 22."
- **What was the date when we spoke, and when is his birthday?**

15 Sean asked his neighbor how old his daughter is. His neighbor replied "My boy is 1/5 the age of my girl, and my girl is 1/5 the age of my wife. My wife is 1/2 as old as I, whereas my grandfather, who is 81, is as old as all of us put together."

- **How old is the daughter?**

16 My mother is now twice as old as me, and she was 31 when I was 8.

- **How old am I now?**

17 How old am I today if my age is 3 times what it will be 3 years from now minus 3 times my age 3 years ago?

18 The average of 3 people's age is 9 years.
- **Find the sum of their age.**

19 Nikki is 13 years old. Her dad, David, is 40 years old.
- **How many years ago was David 4 times as old as Nikki?**

20 Provide the 3 possible scenarios where a father and son's age add up to 66, and, if you reverse the son's age, it is the father's age.

21 Two years ago, Janet was 3 times as old as Adam. In 3 years' time, Adam will be ½ as old as Janet.
- **How old are Janet and Adam now?**

22 When Demi was 6 years old, Shawn was ½ her age. If Demi is 40 now,
- **how old is Shawn?**

23 Al is twice as old as his little sister, Rose. Al is ½ as old as their dad, Mr. Green. After 50 years, Rose will be ½ the age of Mr. Green.
- **How old is Al now?**

24 In 2 years, Shawn's age will be twice his age 5 years ago.
- **How old is he?**

25 If Stormy was born on January 1, 23 B.C. in Abdera and died on January 2, 23 A.D,
- how old was she when she passed away?

26 When Charlie was 31, DJ was 8. Now Charlie is twice as old as DJ.
- How old is DJ?

27 If you take Taylor's age and multiply it by 1.5 times his current age, the product is 24.
- How old is Taylor?

28 One brother says of his younger brother: "2 years ago, I was 3 times as old as my brother was. In 3 years' time, I will be twice as old as my brother."
- How old are they each now?

29 There was a girl ½ my age when I was 12. Now I am 64.
- How old is she?

30

Gino's youth was 1/6 of his life. He grew a moustache after 1/12 more. He married after 1/7 more of his life. He and his wife had a son 5 years later. His son lived 1/2 as long as Gino, and Gino died 4 years after his son.

- **How many years did Gino live?**

31

The ages of a father and son add up to 66.
The father's age is the son's age reversed.

- **How old could they be? (3 possible solutions)**

ANSWERS

1. 12 years old
2. 12 years old
3. 37 years old
4. 8 years old
5. 4 years ago. When Charlotte was 9, her father was 36, 4 times her age.
6. We conversed on January 1st, and her birthday was on December 31st. So, the day before yesterday on Dec. 30th she was 22, and she turned 23 on Dec. 31st. So her next birthday, when she turns 24, would be Dec. 31st of the same year the question was asked. However, next year's birthday would be the following year on Dec. 31st, when she would be 25.
7. They are both 6 years old.
8. Ashley is 22. Her mother is 22 years older, so when Ashley is 22, she's now half her mother's age.
9. 50 years old
10. 41 years ago
11. His daughter is 10 years old, and Akon is 40 years old.
12. Andy is 23, and his father is 46.
13. Maya is 21 years old, and Beanie is 28 years old.
14. His birthday is on December 31, and we spoke on January 1.
15. Five
16. 23
17. 18 years old
18. 27
19. 4 years ago
20. 06 and 60; 24 and 42; 15 and 51
21. Janet is 17. Adam is 7.
22. 37
23. 50 years old
24. 12
25. 45 years old
26. 23
27. 4
28. The elder is 17, the younger 7. Two years ago, they were 15 and 5 respectively, and in 3 years' time, they will be 20 and 10.
29. 58. Half of 12 is 6, so 64-6 is 58.
30. 84 years
31. 51 and 15. 42 and 24. 60 and 06.

01 Tara has $29.00. She bought 4 coloring books that cost $3.00 each and 4 boxes of crayons that cost $2.00 each. She spent the rest of her money on markers.

- **How much money did she spend on markers?**

02 If a custard apple costs $0.80, an apple costs $0.40, and a pomelo costs $0.60,

- **how much will a pomegranate cost?**

03 Mr. and Mrs. Smith were walking home from the shopping mall with their purchases when Mr. Smith began to complain that his load was too heavy. Mrs. Smith turned to her husband and said, "I don't know what you're complaining about because if you gave me one of your parcels, I would have twice as many as you, and if I gave you just one of mine, we would have equal loads."

- **How many parcels was each carrying?**

04 2 mothers and 2 daughters walk into a candy store. Each person buys a candy bar for $0.50 each. They paid $1.50 total for the candy bars.

- **How is this possible?**

05 Getting ready for baseball season, you go to a sporting goods store. A baseball and baseball bat cost $1.10. The ball costs $1.00 less than the bat.

- **How much does the ball cost?**

06 You purchased several items from an online seller. The seller can place items in 8 large boxes or 10 small boxes in a carton. If he was able to ship 96 boxes with more large boxes than small boxes,

- **how many cartons did he send?**

07 Chris, Clara, and Jason have a farm, and they wanted to buy wheat. Jason stayed behind, and Chris and Clara went to the market. Clara bought 75 sacks of wheat, and Adam bought 45. If they split all sacks equally, and Jason paid $1,400.00 for the wheat,

- **how much of the money did Clara and Chris get from the sum?**

08 Erika has $29.00. How much money did she spend on oranges if she bought 4 apples that cost $3.00 each and 4 pineapples that cost $2.00 each?

09

Estelle's 10 friends were all contributing to buy her a baby shower gift. At first, all 10 gave money, but 2 people ended up dropping out and taking their money back. The remaining 8 friends contributed an additional $1 to get the amount back up to how much it originally was before the 2 friends dropped out.

- **How much money were the friends targeting to collect?**

10

Ellie is throwing Malone a surprise party for his 50th birthday. She set a budget for herself on how much to spend for the party. She spent $1/2$ of her money plus $2.00 on the cake. She spent $1/2$ of the remaining money plus another $2.00 on streamers, confetti, and balloons. Then $1/2$ of what was left plus $1.00 was spent on candy, leaving Ellie with no more money.

- **How much was Ellie's budget?**

11

Tamia went to the farmer's market to buy some pomegranates. She bought $1/2$ of the available stock of pomegranate in a stall plus $1/2$ a pomegranate. There was still 1 pomegranate left.

- **How many pomegranates were available in the stall?**

12 If 10 lollipops and 6 bags of chocolate candy cost $0.92, and 6 lollipops and 10 bags of chocolate candy cost $1.00,
- **how much does 1 lollipop cost?**

13 Justin: How much does this bag of scallions weigh?
Shopkeeper: 32 pounds split by ½ its own weight.
- **How many pounds did the scallions weigh?**

14 A bottle of wine and wine bottle stopper cost $2.10, and the wine costs $2.00 more than the bottle stopper.
- **How much does the bottle stopper cost?**

15 Ray, Tamia, Neil, and Ed are roommates. Ray, Neil, and Ed wanted to buy a new shoe rack they saw on sale for $30.00. They gave $30.00 to Tamia since she was on her way to the mall and could make the purchase. When she got to the store, the sales person told her there was a $5.00 discount off the shoe rack, so she made the purchase and left. She wanted to make money off this, so she took $2.00 and gave Ray, Neil, and Ed $3.00; so, in effect, each guy only paid $9.00. Since $9.00x3 is $27.00,
- **what happened to the $1.00?**

16 Ben wants to buy some chocolate bars by the pound. How much does a bar of chocolate weigh if the shopkeeper places 1 bar of chocolate on a pan of a scale and a ³/₄ pound weight and ³/₄ of a chocolate bar on the other and the pans balance?

17 Justin's going back to school in a few weeks, and he needs to buy a notebook. The notebook costs $1 plus ½ its price.
- **How much does the notebook cost?**

18 A belt and a bag cost $150.00. The bag costs $100.00 more than the belt does.
- **How much does each one cost?**

19 Justin went to the pet store and saw that a duck cost $9.00. A spider cost $36.00. A bee cost $27.00.
- **How much will a cat cost?**

20 Selena buys 12 eggs. One the way home, all but 9 crack.
- **How many eggs are left?**

21 Sunday, Melinda and Susan went to the cafe for a cup of tea. The total cost of the bill was $12 and was divided equally among the friends. Melinda paid $4 and Susan paid $4 as well.

- **Who paid the remaining $4?**

22 You have 900ml of milk in a jug, and each cup has a capacity of 100ml. How many cups will you need to distribute the milk evenly between all the cups? But remember you can only fill ¾ of each cup.

23 There is a clothing store in Bartlesville. The owner has devised his own method of pricing items. A vest costs $20, socks cost $25, a tie costs $15, and a blouse costs $30. Using the method,

- **how much would a pair of underwear cost?**

24 An apple plus an apple plus an apple equals 30. An apple plus a bunch of 4 bananas plus a bunch of 4 bananas equals 18. A bunch of 4 bananas minus a whole coconut equals 2.

- **What does half a coconut plus an apple plus a bunch of 3 bananas equal?**

25 Eight bananas cost $0.26. If you had a quarter and a penny,
- **how many bananas could you buy?**

26 You want to purchase some 3-cent stamps. You're at the post office waiting in line for your turn.
- **How many 3-cent stamps will you get in a dozen?**

27 Carly bought dinner for her family of 4 at a fast-food joint. She ordered meal numbers 1, 9, 3, and 7 but returned the food.
- **Why did she do that?**

28 Hayley has $33.00. She bought 4 chocolate bars at $2.00 each, 2 notebooks at $4.50 each, and spent the rest of her money on sneakers.
- **How much did the sneakers cost?**

29 Hayley went to buy candy. She had 1 penny, 2 quarters, 3 nickels, and 8 dimes. The candy cost $0.95. She planned on bringing back 3 coins to put in her piggy bank.
- **Which coins did she use to buy the candy?**

30 April had $28.75. She bought 3 biscuits that cost $1.50 each, 5 magazines that cost $0.50 each, 5 watermelons for $1.25 each, and used the rest of the money to buy a hat.

- **How much was the hat?**

31 In an odd little town, there was an odd little stream with odd little fish in an odd little team. A stranger approached a local fisherman and asked him how much his odd little fish weighed. The odd little man replied, "All the fish in this stream weigh exactly ½ of a pound plus ½ of a fish." Isn't that odd?

- **How many pounds does an odd little fish weigh?**

32 Katherine went to the bakery to buy bread. Her mother gave her money but made her promise to bring back one coin when she returned home. Katherine had 2 pennies, 2 quarters, 3 nickels, and 4 dimes. The bread cost $0.82.

- **Which coin did she bring back to her mother?**

33 If 10 bags of jelly beans and 6 licorice sticks cost $1, and 10 licorice sticks and 6 jelly bean bags cost 92 cents, then

- **how much does one licorice stick cost?**

34

A man has been selling tomatoes in a local supermarket. He has taken 100 boxes of tomatoes with him and has previously marked all these boxes with numbers 1 to 100.

- **How many times has the man written the number 8?**

35

Farmer Brown came to town with some watermelons. He sold ½ of them plus ½ a melon and found that he had one whole melon left.

- **How many melons did he take to town?**

36

"How much is this bag of potatoes?" asked the man. "32lb divided by ½ its own weight," said the grocer.

- **How much did the potatoes weigh?**

37

A little boy goes shopping and purchases 12 tomatoes. On the way home, all but 9 get mushed and ruined.

- **How many tomatoes are left in a good condition?**

38

How much water is added to 750g milk to get 1 kilogram of liquid?

39 A farmer is selling hens on the local market. 2 hens can lay 2 eggs in 2 minutes. If this is the maximum speed possible,

- **what is the total number of hens needed to get 500 eggs in 500 minutes?**

40 A fisherman is selling fried fish in an open local market. He is frying them in a pan which can only hold 2 fish at one time. It takes 5 minutes to fry 1 side of the fish.

- **What is the shortest time in which he can fry 3 fish in the same pen?**

41 A man buys a horse for $60. He sells the horse for $70. He then buys the horse back for $80. And he sells the horse again for $90. In the end,

- **how much money did the man make or lose? Or did he break even?**

42 Irene went to the store to get ice cream. She had 2 quarters, 8 dimes, 3 nickels and 1 penny. The ice cream cost her $0.95. She promised not to spend 3 of her coins.

- **Which coins did she use to buy the ice cream?**

43 A farmer in Australia grows a beautiful pear tree, which he harvests to supply fruit to all the nearby grocery stores. One of the store owners has called the farmer to see how much fruit is available that he can buy. Unfortunately, the farmer isn't currently near the tree, so he has to work it out in his head. He knows that the main trunk of the tree has 24 branches, that each branch has 12 boughs, and that each bough has only got 6 twigs. Each one of these twigs bears one piece of fruit,

- **so how many plums will he be able to sell to the store owner?**

44 Andy bought a 100-page book from a bookstore but then realized that ⁴/₅ of the book was missing.

- **How many pages were there in the book when Andy bought it?**

1. $9.00
2. The price is determined by multiplying $0.20 by the number of vowels; therefore, the price of the pomegranate will be $1.00.
3. Mrs. Smith was carrying 7 parcels, and Mr. Smith was carrying 5.
4. It was a daughter, a mother, and a grandmother that walked into the candy store.
5. The baseball bat costs $1.05, and the baseball costs $0.05.
6. 11 cartons (56 boxes (7 large boxes) +40 boxes (4 small boxes)).
7. Chris: $175.00 and Clara: $1,225.00
8. $9.00
9. $40.00
10. $20.00
11. 3
12. $0.05
13. 8 pounds
14. $0.05
15. The shoe rack cost $25.00. Add the $3.00 the 3 guys received making it $28.00. Then the $2.00 Tamia kept for herself tallies the amount to $30.00.
16. 3 pounds
17. $2.00
18. The belt costs $25.00, and the bag costs $125.00.
19. $18.00
20. 9
21. Their friend Sunday
22. You will need 12 cups. The total capacity of each cup is 100ml. But ¾ of 100ml

is 75ml. Now divide 900ml by 75ml. It comes out to be 12. Therefore, you will need 12 cups of 100ml each to distribute 900ml milk equally, when you can only fill ¾ of each cup.

23. $45. The pricing method consists of charging $5 for each letter required to spell the item.
24. 14
25. Eight
26. 12
27. Because the food was odd.
28. $16.00
29. The 3 nickels and 8 dimes.
30. $15.50
31. 1
32. One quarter
33. The answer is 5 cents. Jelly bean bags cost 7 cents.
34. The answer is 20! 8, 18, 28, 38, 48, 58, 68, 78, 80, 81, 82, 83, 84, 85, 86, 87, 88, 89, 98.
35. 3 melons
36. 8 lbs
37. 9 tomatoes
38. 0.25 kg
39. 2 hens
40. He puts 2 fish in the pan. He fries them for 5 minutes and then takes out 1 and turns the other 1. He also puts in the third fish. Now he fries those for 5 minutes. The fish that he turned must be ready now, so he will take it off while turning the other one. He will then put in the fish which was fried on one side only in the pan. Then he will fry them for 5 minutes, and all 3 fish are fried now.
41. The man made $20.
42. The 8 dimes and 3 nickels.
43. None! He doesn't own a plum tree…he owns a pear tree!
44. 20 pages

01 How do you change 98 to 720 by using just one letter?

02 How many 9s are between 0 and 100?

03 If a mother and father have 4 daughters, and each one of those daughters has a brother,
- **how many people are in the family?**

04 What did the calculus book say to the biology book?

05 How much does one brick weigh if one brick is ½ a brick and one kilogram heavy?

06 Make a true equation with the symbols = and + and the numbers 5, 4, 3, and 2

07 What number is not larger than 2 times that number when multiplied by 3?

08 If you have 14 black socks, 14 brown socks, and 14 blue socks in a drawer, without looking,
- **how many socks should you take out to ensure you have a matching pair?**

09 If you are treating your friends to a movie, which would save you more money: taking 2 friends at the same time, or taking 1 to the movies twice?

10 How many days are in 4 years?

11 What is $9/10$ of $1/2$ of $8/9$ of $2/3$ of $7/8$ of $3/4$ of $7/8$ of $4/5$ of $6/7$ of $5/6$ of 1,000?

12 What do math professors like to eat?

13 How do you, without using a 1-dollar bill, give someone 7 bills that total $83?

14 How tall is Mr. Brown's wife if she is ½ his height plus 90 centimeters?

15 What do you get when you divide 20 by ½ and then add 30?

16 If LV blows 18 bubbles, pops 5, eats 6, pops 7, and blows 1, how many bubbles are left?

17 A point has 0, a sphere has three, and a circle has 2. What are we talking about?

18 What was remarkable about 3,661 seconds after midnight on January 1, 2001?

19 What is significant about 12:34PM, May 6, 1978?

20 Using five 2s and plus signs, how do you get 28 as the sum?

21 If 6 dogecoins are worth 1 ½ bitcoins,
- how many bitcoins are 27 dogecoins worth?

22 How many plus signs are needed, and where do you put them, to get a sum of 99 from 987,654, 321?

23 If a building has 6 stories that are all the same height,
- how long is the climb up to the 6th floor compared to the climb up to the 3rd floor?

24 Ellie had marbles that she wanted to arrange in a solid square. With her first arrangement, she had 39 extra marbles. When she added the number of marbles on a side by one, she discovered that she would need 50 more marbles to make a new square.
- How many marbles did Ellie have?

25 In this sequence of numbers, what comes next?
- 2, 3, 5, 9, 17

26 A train leaves Buffalo for Manhattan at 60 miles per hour. Another train leaves Manhattan for Buffalo at 40 miles per hour.

- **What is the distance between the 2 trains 1 hour before they pass each other?**

27 A snowflake has how many sides?

28 Juan and Carlos were both pushing wheelbarrows with bricks in them. If Juan gives Carlos 1 of his bricks, Carlos will have twice as many bricks as Juan. If Carlos gives Juan 1 of his bricks, they would have an equal number of bricks.

- **How many bricks do they each have?**

29 You toss a die 9 times and get a 5 all 9 times.

- **What are the chances you'll get another 5 on the next toss?**

30 How many snakes do you need to kill 100 rats in 50 minutes if 7 snakes can kill 7 rats in 7 minutes?

31 Adam was given 3 pills by a doctor and told to take 1 every 30 minutes.
- **How long will the pills last?**

32 If Raymond wrote numbers 300 to 400 on the blackboard,
- **how many times did he write the number 3?**

33 Jessie and Ellie have 6 sons, and each son has a sister.
- **How many people are in the family?**

34 Compute ½ of ¼ of 2/9 of 3/7 of 84?

35 In a day and a half, a turtle and a half can lay an egg and a half. In half a dozen days, half a dozen turtles can lay
- **how many eggs?**

36 How many times does the digit 5 occur between 1 to 100?

37 What is the answer to the question mark?
1=4
2=16
3=64
4=?

38 Lionel's car has a mileage of 72,927.
- **How far does he need to travel in order to form another palindrome on his car's odometer?**

39 When Lourdes died, she left ½ of her money to Bruno. ½ of the amount Bruno got went to Janet. ⅙ went to Adam. The rest of the money, $1,000.00, went to the dog shelter.
- **How much money did Lourdes leave total?**

40 In Harrisburg, the mayor wished to have more women than men in the village. He declared that all child-bearing couples must continue bearing children until they have a girl but must stop having children once they bear a daughter.
- **What is the expected ratio of boys to girls in Harrisburg?**

41 If 2 bears weigh 120 kilos, a rabbit and a bear weigh 70 kilos, and a bear, fox, and rabbit weigh 90 kilos,
- how much does each animal weigh?

42 Which weighs more: 16 ounces of salad dressing or a pound of turkey?

43 Nathan bought a cell phone and a cell phone case for $110.00. The cell phone costs $100.00 more than the case.
- How much does the cell phone cost?

44 If muffins are $0.12 a dozen,
- how many muffins can you get for $1.00?

45 If Fred has 2 kids and the eldest is a boy,
- what are the chances that the next one is also a boy?

46 What does $4k+2k+k$ equal?

47 There are a number of case files on the shelf. Selena took the case file that is 6th from the left and 4th from the right.
- **How many case files are on the shelf?**

48 If the 7 dwarves were brothers, were all 2 years apart in age, and the youngest was 7 years old,
- **how old is the eldest dwarf?**

49 A villager asked a farmer how much his watermelon weighed. The farmer said the watermelon weighed ½ a pound plus ½ of a watermelon.
- **How much does the watermelon weigh?**

50 If 2 turtles can lay 2 eggs in 2 minutes,
- **how many turtles do you need if you need 300 eggs in 300 minutes?**

51 During a military exercise, the soldiers were facing west. Their commander shouts "Right turn. About face. Left turn."
- **What is their orientation now?**

52 If 4 people can build 4 houses in 4 days,
- how many houses can 8 people build in 8 days?

53 Francis weighs ½ as much as Chito. Luca weighs 3 times as much as Francis. They weigh 720 pounds total.
- How much does each person weigh?

54 If ½ of 5 is 3, what is ⅓ of 10?

55 Ice is very claustrophobic. She needs to get on a train that goes in a tunnel as soon as it leaves the station.
- Where is the best place to sit if you were her?

56 Mr. Brown can make a cigarette from 4 used butts. He was able to collect 16 butts.
- How many cigarettes can he make?

57 How can you make 0.9=1?

ANSWERS

1. Add 'x' to 90 and 8. 90 x 8 = 720.
2. 20
3. 7
4. "Man, I have so many problems."
5. One brick weighs 2 kilos.
6. 4+3=5+2
7. 0
8. 4 socks
9. It would be cheaper to take 2 friends at the same time, because you would just be buying 3 tickets compared to 4 tickets if you take 1 friend twice.
10. There are 1,461 days in 4 years ((365 days x 4 years)+1 (due to the leap year)).
11. 87.5
12. Pi
13. Give them four $2.00 bills, one $50.00 bill, one $5.00 bill, and one $20.00 bill.
14. 180 centimeters
15. 70. [20/½ = 40 +30 = 70]
16. 1
17. Dimensions
18. It was 01/01/01 01:01:01
19. The time and date are 12:34 5/6/78.
20. 2+2+2+22=28
21. 6 ¾ bitcoins
22. There are 2 possible answers:
    Six plus signs
    9+8+7+6+5+43+21=99
    Seven plus signs
    9+8+7+65+4+3+2+1=99
23. 2 ½ times as long.
24. Ellie has 1,975 marbles.
25. 33
26. 100 miles (60+40)

27. 6
28. Juan had 5 bricks, and Carlos had 7.
29. One out of 6. Previous tosses have no impact on a new toss.
30. 14
31. One hour
32. 120 times
33. 9 people
34. 1
35. 24 eggs
36. 20
37. 256
38. 110 miles
39. $12,000.00
40. 1:1
41. The bear weighs 60 kilos. The rabbit weighs 10 kilos. The fox weighs 20 kilos.
42. Neither — they both weigh the same.
43. $105.00
44. 100 muffins
45. 50 percent
46. 7,000
47. 9 case files
48. 19 years old
49. One pound
50. Two turtles
51. East
52. 16 houses
53. Francis weighs 120 pounds, Chito weighs 240 pounds, and Luca weighs 360 pounds.
54. 4
55. The back of the train since it will be going faster as it accelerates.
56. 5 cigarettes. He can make 4 cigarettes, and when he's done smoking them, he can make one more.
57. There are 2 possible answers:
    $0.9 \neq 1$
    $0.\underline{9} = 1$

01 In a marathon, if Al came in 2 places before the last man and finished 1 ahead of the man who finished 5th,
- how many participants were there?

02 Reign has as many brothers as she has sisters. Each of her brothers has double as many sisters as brothers.
- How many daughters and sons does Reign's father have?

03 How can you take 5 matchsticks to make 8?

04 If there are 25 blue balls, 47 red balls, and 3 green balls in a container, and you were blindfolded,
- what is the minimum number of balls that you have to pick to make sure you have at least 2 balls that are of different colors?

05 Mr. Wolf has 4 sons, and each of his sons has a sister.
- Mr. Wolf has how many children?

06 How can you add 8 number 8s to get 1,000?

07 Khiry, Hakeem, and TJ are brothers. Their current age is prime.
- **How old are they if the difference between their ages is also prime?**

08 Michael has $1.19 in the form of 10 coins. He is not able to make exact change for a nickel, a dollar, dime, half-dollar, or quarter with these coins.
- **What coins does he have?**

09 Add 5 to 6 and I get 11. Add 6 to 7 and I get 1.
- **What am I?**

10 When are 1600 minus 40 and 1500 plus 20 the same thing?

11 How do you make 3 equal squares out of 8 sticks so that 4 are twice as long as the length of the other 4?

12 You want to boil an egg for 2 minutes. You have a 5-minute timer, 4-minute timer, and 3-minute timer.

- **How do you use these to boil the egg for 2 minutes?**

13 What number will come next?
- **6,457; 4,576; 5,764; ...**

14 At a certain department store's storage area, there are 600 mannequins. 5% of these mannequins have 1 arm. Of the other 95% of the mannequins, $1/2$ have 2 arms, while the other $1/2$ do not have any arms.

- **How many arms are there in total?**

15 Haley is the head organizer for the World Chess Championships this year. There are 657 countries participating. The rules of the tournament are set up so that whoever loses a match is eliminated, and whoever wins advances to the next match. Due to the uneven number of participants, the first match leaves 1 player out, and they automatically advance to the next round.

- **How many matches does Haley need to have to determine who the 2020 champion will be?**

16 8 years ago, Barbara was 8 times the age of her son, Mason. Today, if you add their ages, the sum is 52.
- **How old are Barbara and Mason?**

17 Erika and Eric bet each other pizza for every golf match they win against each other every week. The pizza is purchased at the end of each week. The same number of wins means the pizzas get cancelled. So if Erika won 4 matches and no pizzas and Eric won 3 pizzas,
- **how many golf matches did they play?**

18 Ben, a student driver, is practicing driving around a track where there are 12 cones that are equidistant on the track. Ben starts at the first cone and reaches the 8th cone 8 seconds after he starts. If his speed is constant,
- **how many seconds does he need to reach the 12th cone?**

19 How many sisters and brothers are there in the family if a girl has as many sisters as brothers, but each brother only has ½ as many sisters as brothers?

20

Two cornmeal dealers wish to sell their barrels of cornmeal at the market in Cameroon. One dealer has 20 barrels of cornmeal, and the other has 64 barrels. Upon arriving, they discover they do not have enough money to pay the tax for the cornmeal, so the first one pays in 2 barrels of cornmeal and receives 40 francs as change. The second dealer pays 40 francs and 5 barrels of cornmeal.

- **How much does each barrel of cornmeal cost, and how much is the tax?**

21

A toolbox contains nuts and bolts. If the bolts make up 25% of the number of items in the box,

- **what percentage of the entire toolbox is the bolts?**

22

How can you turn five 1s into 100? (You can use +, -, x, /, parentheses, and brackets.)

23

How can you turn five 5s into 100? (You can use +, -, x, /, parentheses, and brackets.)

24

How can you turn five 3s into 37?

25 How can you turn five 4s into 55?

26 How can you turn four 9s into 20?

27 Joe found himself lost in the woods. He stumbled upon 2 campers a few days later. One of the campers, Bill, had 3 fish and the other camper, Chris, had 4 fish. All the fish were pretty much the same size. Chris and Bill decided to pool all their fish together and the 3, Joe, Bill, and Chris, would eat the fish together. Once they were done eating, Joe paid Bill and Chris $8.00 as payment for the fish.

- **How should Bill and Chris split the money fairly?**

28 How can a glass ball the size of a ping pong ball avoid being crushed by a cast iron ball with a diameter of 300 millimeters that is circling around the room?

29 Sam lives at the reverse of Kevin's house number. The last number of the difference between their house numbers is 2.

- **What are the lowest possible numbers of their houses?**

30 Nathan has 8 blocks that are all the same size. 7 of the blocks weigh the same, while 1 weighs a little heavier than the others. There is a balance scale available that Nathan can use only twice to find out which block is the heaviest.

- **How does he do this?**

31 Erica and Nick are about to have a paintball war. Erica doesn't think this is a fair fight, because Nick has 3 times as many paintballs as she does. Nick gives her 10 paintballs, but she still doesn't think it is a fair fight because Nick has twice as many paintballs as she does.

- **How many paintballs does Nick need to give Erica for the fight to be fair?**

32 Taylor needs to get 8 pieces from her daughter's round birthday cake so everyone can get a slice. Since her daughter is 3 years old, tradition dictates that only 3 cuts can be made to the cake.

- **How can she get 8 slices with just 3 cuts?**

33 If in Tunisia, half of $10.00 is DT 6.00,

- **what would 1/6 of $30.00 be?**

34 What does the question mark equal?
A+B=76
A-B=38
A/B=?

35 What does the question mark equal?
1+4=5
2+5=12
3+6=21
8+11=?

36 Jason found 5 chains made of 4 gold links. He would like to combine them into 1 loop of 20 gold links to make a necklace. He brought it to the jeweler, and he was told the cost will be $10 per gold link that has to be broken and resealed.

- **How much will it cost to make the necklace?**

37 Your TV remote uses 2 working batteries. Out of 8 batteries, only 4 are working.

- **What is the fewest number of tries you must do to make sure you can get the remote to work?**

38

Kesha and her son, Sean, live a short car ride away from Sean's school. Every school day, Kesha gets in her car at the same time every afternoon, drives to school, and picks up Sean when swim practice ends — exactly at 5PM, and they return home. One day, Sean isn't feeling well so he leaves swim practice early and starts walking home. After Sean has been walking for an hour, Kesha, on her usual route to pick him up, comes across him and they go home together arriving 40 minutes earlier than normal.

- **How much swim practice did Sean miss?**

39

Taylor has 2 containers with her. 1 container has blue balls and the other has orange balls. These 2 containers have an equal number of balls.

- **How can she increase the possibility of pulling a blue ball from each of the containers?**

40

Katie had a belt that was so long it could wrap around the circumference of Earth. Her friend, Manny, decides to one-up her and shows her his belt that is a little bit longer than hers by exactly his height: 6 feet. If Manny can also wrap his belt around Earth,

- **how far above the Earth can he pull the belt if he pulled it uniformly and tautly?**

41 A lizard is at the bottom of a 20-meter tall wall. Every day the lizard climbs 5 meters up the wall and slides back down for 4 meters.

- **How long will it take the lizard before it reaches the top of the wall?**

42 A motorcycle rider is traveling south at 25 kilometers per hour. A train going south passes the motorcycle at 40 kilometers per hour. The motorcycle rider passes a jogger going south at 5 kilometers an hour.

- **Who will be moving further away from the motorcycle at a faster pace, the jogger or the train?**

43 You have a barrel full of wine, and you need to get 1 gallon. You only have a 3-gallon carafe and a 5-gallon carafe.

- **How do you measure out the 1 gallon you need?**

44 A 100-pound cantaloupe was left out in the sun. 99% of the cantaloupe's weight is water. After a couple of hours under the sun, 98% of the cantaloupe's weight is water.

- **How much water evaporated?**

45 What is the angle if the hour and minute hands are pointing at 3:15?

46 Foot rugs are made from pieces of cloth in a factory. Each piece of cloth is enough for 1-foot rug. Cutout pieces collected for making 6-foot rugs can be gathered and made into a piece of cloth.

- How many foot rugs can be made from 36 pieces of cloth?

47 On a 100-seater airplane, 100 passengers start boarding. If the first passenger gets on the plane and randomly picks a seat, and the remaining passengers either take their assigned seat if it is available or randomly pick one if it is taken,

- what are the chances the 100th passenger will get to sit in his assigned seat?

48 A right cylinder-shaped glass looks like it is about half full with juice.

- How can you make sure, by just using the glass, if it is half full or not?

49

2 cables will take 1 hour to burn from 1 end to the other.

- **How can you burn both in 45 minutes without cutting the cables?**

50

The first sum-day of the 21st century is on January 1, 2002. A sum-day is the date where the day and month add up to the year, so in this case 1+1=2. When is the last sum-day of the 21st century?

- **There are how many sum-days in the 21st century?**

51

Rob owns a toaster that has 2 slots. This toaster is able to toast only 1 side of each piece at a time. This takes 1 minute. If Rob wants 3 pieces of toast,

- **how fast can he toast all 3 slices on both sides?**

52

Hanson and Michelle are running a 100-meter race. Hanson wins the first race by 5 meters. For the second race, Hanson stands 5 meters behind the starting line. Assuming the same speed is kept as in the first race,

- **who will win the race?**

53 You need to time 15 minutes but only have a 13-minute and 11-minute hourglass.

- What do you do?

54 In the middle of the night, Dan, John, Jacob, and Rob need to cross a bridge that has railroad tracks over it. The train is scheduled to arrive, and the 4 must cross the bridge in 17 minutes. The friends only have 1 flashlight, and it needs to be carried back every time it is carried to the other side of the bridge. Dan can cross the bridge in 1 minute, John in 2 minutes, Jacob in 5 minutes, and Rob in 10 minutes. If 2 people are crossing the bridge together, they can only go as fast as the slowest person.

- How can they cross the bridge in time?

ANSWERS

1. 6 participants
2. Each brother has 4 sisters and 2 brothers, and each daughter has 3 brothers and 3 sisters, so Reign's father has 3 sons and 4 daughters.
3. Arrange the matchsticks to form a roman numeral 8.
4. 48 balls since there is a slight chance you pick up 47 red balls in a row.
5. 5 children
6. 8+8+8+88+888=1,000
7. Khiry is 5, Hakeem is 2, and TJ is 7.
8. 4 pennies, a half-dollar, 4 dimes, and a quarter
9. A clock
10. Military time
11. The 4 longer sticks should be used to be the ones sharing sides between the squares and the shorter sticks fill in the rest of the squares. There should be 3 interlaced squares.
12. Once the water in the pot boils, turn the 5-minute and 3-minute timers over. When time runs out on the 3-minute timer, put the egg in the pot. When time runs out on the 5-minute timer, it means 2 minutes has passed.
13. 7,645. The first digit is moved to the back to make the next number.
14. 600
15. 656 matches
16. Barbara is 40, and Mason is 12.
17. 11 matches. Eric won 7 of the 11 matches to win the 3 pizzas and cancel out the 4 pizzas Erika won.
18. 12 ½ seconds
19. 4 sisters and 3 brothers
20. A barrel of cornmeal costs

120 francs, and tax is 10 francs for each barrel.
21. 20%
22. 111-11=100
23. There are 3 possible answers:
    (5x5x5)-(5x5)=100
    (5+5+5+5)x5=100
    (5x5)(5-(5/5))=100
24. There are 3 possible answers:
    33+3+3/3=37
    333/3x3=37
    3x3x3+3/.3=37
25. 44+44/4=55
26. 9+99/9=20
27. Bill should get $1.00, and Chris should get $7.00.
28. Since the cast iron ball is more than 5.83 times the diameter of the glass ball, the glass ball just needs to hug the wall to be safe from being crushed by the cast iron ball.
29. 91 and 19
30. Split the blocks into 3 groups: 2 groups consisting of 3 blocks, and 1 group consisting of 2 blocks. Weigh the 2 groups of 3 blocks against each other. If they balance, it means the heavy block is in the group of 2 blocks and you would just need to weigh those 2 blocks against each other to see which one is heavier.

    If the group of 3 blocks do not balance, weigh 2 of the 3 blocks. If they balance, the heavier block is the unweighed one. If they do not balance, just weigh the blocks against each other to see which is heavier.
31. Nick must give Erica 20 more paintballs so each of them will have 60 paintballs.
32. Make a parallel cut through the middle of the cake. Then make a crosscut perpendicularly through the cake (making these cuts 2 and 3).
33. DT 6.00
34. 3
35. 96
36. $40 if you break 4 links in 1 of the chains and use those to attach the other 4 chains.
37. 7 tries
38. Sean missed an hour and 20 minutes of swim practice.

39. Put a blue ball in one container and the rest of the balls in the other container increasing the chances of pulling a blue ball from each container to 75%.
40. 0.95 feet
41. 16 days
42. The jogger
43. Fill the 3-gallon carafe and pour the contents into the 5-gallon carafe. Fill the 3-gallon carafe again and pour out the contents into the 5-gallon carafe until it's filled to the brim. The remaining wine in the 3-gallon carafe is 1 gallon.
44. 50 pounds
45. 7.5°
46. 43
47. 50%
48. If the juice lines up perfectly with the bottom rim of the glass when you tip it, it means it is half full.
49. Burn both ends of one cable at the same time while also burning one end of the other cable. The cable with both ends burned will be completely burned and the other one will be half burned in half an hour. Light the unlit side of the remaining cable, and it will be completely burned in 15 minutes, making it 45 minutes total.
50. The last sum-day is on December 31, 2043. There are 365 sum-days in the 21st century.
51. 3 minutes
52. Hanson
53. Flip over both timers. When the 11-minute one is done, flip it over immediately. When the 13-minute on runs out, flip over the 11-minute hourglass again so it will time another 2-minutes.
54. Dan and John will cross first (2 minutes). Dan will bring the flashlight back (1 minute). Jacob and Rob will cross next (10 minutes). John will bring back the flashlight (2 minutes) and both John and Dan will cross to the other side (2 minutes).

01

A pollster needed to find out how many children live in the village along with their ages. He went to the first house, and a woman answered the door. He asked her if she had children, how many there were, and their ages. Instead of answering the question, she provided hints:

- **Hint 1:** The product of the 3 kids' ages is 36.
  The surveyor told her that information was not enough, so the woman provided a second clue.
- **Hint 2:** The sum of the kids' ages is the number of the house next door.
  The surveyor went to the house next door, looked at the house number, went back to the woman, and told her that the information was not enough to get an answer. The woman provided a final clue.
- **Hint 3:** The oldest kid plays the piano.

02

Should you accept the bet if someone tells you that they will bet you $1 that if you give them $2, they will give you $3 in return?

03

The sun is 100 million miles from Earth. It takes 8 minutes for light to reach Earth from the sun, and the speed of light is 186,000 miles per second. If the sun rose at 6AM today, and the speed of light doubled its normal speed,

- **what time will the sun rise tomorrow?**

04
If 6+5=111, 8+1=79, and 4+2=26, what is 7+3?

05
Elvis and Marshall were on the way to the countryside via train from the city for a vacation. Elvis noticed that the trains coming from the countryside passed them every 5 minutes. "How many trains from the countryside arrive in the city in 1 hour if the train runs at equal speeds in both directions?" Elvis asked Marshall. Marshall answered, "12 because 60 minutes divided by 5 minutes is 12."

- **Is this correct?**

06
You are walking on train tracks that run on a narrow bridge that is over a 150-foot drop into rocks. You're 3/8 of the way across when you hear the train whistle behind you. Running at 10 miles an hour, either way you go, you can reach safety right before getting run over. In this case,

- **how fast is the train going?**

07
If 100 Easter eggs are placed in a straight line, 1 foot apart from each other in a garden, how many feet must someone who needs to pick them up walk to place them in a basket that is 1 foot away from the first Easter egg?

08 A recipe calls for baking apple and orange bread for 9 minutes in the oven. You only have a 7-minute hourglass and a 4-minute hourglass.

- **How are you going to time the apple and orange bread so it bakes for exactly 9 minutes?**

09 Paula had a huge 40-pound bar of chocolate that she divided into 4 portions so that she can use them as weights on a balance scale and weigh anything 40 pounds and below.

- **What is the weight of each piece of chocolate?**

10 How can you arrange the numbers 1-9 into the equation below to get 100?
(-) + + - - - = 100

11 Clifford only had $2.00, but the bus fare home costs $3.00. He went to a pawnshop and pawned his $2.00 for $1.50. He then went to his friend Mary and made a deal to sell the $2.00 pawn ticket to her for $1.50, and she agreed. Now Clifford has $3.00.

- **Who lost a dollar and why?**

12 Two roommates who were also cyclists went on a ride, and one of them brought their pet dog with them. One cyclist started from Nairobi while the other one started from Mbale. When the cyclists were 180 miles apart, the dog got loose and ran ahead to meet the other cyclist. After the dog reached the cyclist, he turned around to get back to the other cyclist. The dog kept running back and forth between the 2 cyclists until both cyclists met. The cyclists were biking at 15 miles per hour, and the dog was running at a speed of 30 miles per hour.

- **How many miles did the dog run?**

13 Using the number keypad on your computer's keyboard, how many 10-digit number combinations can you make using the digits 0 to 9 only once and using the knight's move (L-shape move) used in chess to move from one key to the next?

14 There are 12 balls that are all the same size. 11 of these balls are the same weight, 1 is a different weight, but you were not told if this is lighter or heavier. You are given a balance scale that you may only use 3 times to determine which ball is the 1 of a different weight and determine whether this ball is lighter or heavier than the others.

15 Jess will give DJ $1,000 if, given 10 boxes and $1,000.00 in $1 bills, he can place the money in the boxes in such a way that no matter what number of dollars Jess asks for, DJ can give Jess the exact amount she is asking for without having to take out money in any of the boxes and transfer to a different box.

- **How should the money be distributed among the boxes?**

16 Candy was being given away to some children. When the candy is evenly distributed to 4 children, there are 3 pieces of candy leftover. If the candy is distributed to 3 children, 2 pieces of candy remain. If the candy is evenly distributed among 5 children, 2 pieces of candy remain.

- **How many pieces of candy are there?**

17 Using the digits below once each, make 2 fractions so that when added together, the sum equals 1.
0, 1, 2, 3, 4, 5, 6, 7, 8, 9

18 Michelle is 21 years older than her daughter. In 6 years, Michelle will be 5x older than her daughter.

- **Where is the father?**

19 Sitti had a farm with only chickens, cows, and pigs. Her neighbor asked her how many of each animal she had on her farm, and Sitti said, "All are chickens except 3, all are cows except 4, and all are pigs except 5."

- How many of each animal is on Sitti's farm?

20 How do you make 3 zeros equal to 6?

21 Adam wants to send David his priceless antique guitar pick through the mail but wants to make sure it doesn't get stolen. The only way it won't get stolen is if Adam puts the pick in a box and puts a lock on the box. Both Adam and David have locks and keys but not the same ones.

- How can the guitar pick be sent safely through the mail?

22 Tamia, Howard, and Jody played a round-robin tennis tournament. The winner stayed on and played the person who sat out the previous game. At the end of their tournament, Howard played the last 7 games. Tamia played 8 games. Howard played 12 games. Jody played 14 games.

- Who played the 4th game and who won?

23

Chris and Sean are identical twins. They decide to race each other one day. They will start at the bottom corner of their respective cuboid platforms to reach the bell they will ring, which is located at the opposite corner. The dimensions of Chris's platform are 3 feet high, 3 feet long, and 3 feet wide. The dimensions of Sean's platform are 2 feet high, 4 feet long, and 3 feet wide. If they both run at the same speed and take the shortest possible route to reach the bell,

- **who will reach the bell first?**

24

Bruno and Curtis are facing off in a contest involving 100 silver pieces and 1 gold bar. Bruno and Curtis will take turns taking at least 1 but no more than 5 pieces of silver from the pile. Whoever gets the last piece of silver will also get the gold bar. Bruno wins the coin toss and gets to decide who goes first.

- **Who should he choose so he can make sure he gets the gold bar?**

25

In a series of numbers:
9 8 7 6 5 4 3 2 1
Make 100 by using any number of minuses and pluses. However, you cannot change the order of the numbers.

- **What is the least number of minuses and pluses you need to make 100?**

26

Five siblings are left with an inheritance of 100 gold pieces. They can decide how the gold is divided based on a vote. Each sibling can propose a plan on how the inheritance will be split. If a minimum of 50% agrees on the proposed plan, that will be the plan that will be followed. If less than 50% agree, the sibling who proposed the plan will be cut out of the will.

- **What plan can the eldest sibling come up with to get as much of the inheritance as possible?**

27

In the upcoming school election, 5 students are running for president: Taylor, Anna, Nikki, Leighton, and Hayley. In this school, there are always 4 consecutive elections held for president. 2 candidates run against each other, and the 1 who loses gets eliminated. The 1 who wins runs against the next candidate, and so on, until we get to the final 2 candidates, which will determine who will win as president. Leighton isn't feeling too good about winning because she has been consistently ranked at the bottom by polls. She does know that there are 5 groups in the school that have the following voting preferences:

Group 1: Anna > Taylor > Nikki > Leighton > Hayley
Group 2: Anna > Taylor > Leighton > Hayley > Nikki
Group 3: Hayley > Nikki > Anna > Taylor > Leighton
Group 4: Taylor > Hayley > Nikki > Anna > Leighton
Group 5: Taylor > Nikki > Anna > Leighton > Hayley
Leighton, by a stroke of luck, was asked to pick the voting order.

- **What order should she choose?**

28 There are 2 cats in front of 2 other cats. And there are 2 cats behind 2 other cats. There are 2 cats beside 2 other cats.
- How many cats are there?

29 Determine the pattern:
2+2=44
3+3=96
4+4=168
5+5=2,510
6+6=?

30 What are your chances of getting the answer right if you randomly choose any of the answers below?
a. 50%
b. 25%
c. 0%
d. 25%

31 Dharma was given $300.00 to bring home to her mother. While she was walking home, Mr. Brown stopped her and told her she could double her money. He will give her $600.00 if she can roll a die and get a 4 or higher, or she can roll 2 dice and get a 5 or a 6 on at least one die, or she can roll 3 dice and get a 6 on at least one die. If she doesn't, Mr. Brown will take her $300.00.
- What should Dharma do?

32

Once the quarantine was lifted, 2 friends, Manny and Chris, headed out to the local bar. They found 4 other couples at the bar. The people who have never met before all shake hands with each other. Chris found out that in this group of 10 people, there were many who had known each other before, and when he asked everyone how many hands they shook, he got 9 different answers.

- **How many hands did Manny shake?**

33

Ty deals cards for himself and his friend Anderson. He does not deal the same number of cards. If Anderson gives Ty some cards, Ty will have 4 times as many as Anderson. If Ty gives Anderson the same number of cards, he will hold 3 times as many cards as Anderson.

- **How many cards do they trade, and how many cards does each one have?**

34

Joe believes he has so much he needs to do that there is not enough time to go to school, his argument being he needs 8 hours of sleep a day, which is 122 days a year. Weekends are 104 days a year, and summer vacation is 60 days a year. Meals are 3 hours a day, translating to 45 days a year. Add another 30 days or 2 hours per day for exercise. Adding it all up, that's already 361 days, leaving only 4 days out of the year for school.

- **Where is the flaw in his argument?**

35

Rob went to the city with $10.00 in his pocket. He bought a book for $6.50, then bought a hot dog and coffee for $3.00. He took the train with his remaining $0.50, which would take him 7 stops. He had to get off 5 stops away from where he got on though.

- **What could the reason be?**

36

Mrs. Keys harvested 150 apples from her tree and asked her 3 sons, Bob, Mark, and Bruno, to take them to the market and sell them in a way that they each bring home the same amount of money. She then gave Bob 15 apples, Mark 50 apples, and Bruno 85 apples.

- **How did the boys accomplish this?**

37

A coin that's heavier on one side comes up heads more than tails. Dave and Hanson each want to keep the coin so they will have a coin toss using this coin to see who will win it.

- **Is there any way to do this fairly?**

38

John has to plant 4 bushes that are all the same distance from each other.

- **How does he do this without any measuring tools?**

39 Rob has 30 friends coming over. One of his friends asks for ½ a cup of water. A second friend asks for ¼ of a cup of water. The third one asks for ⅛ of a cup, and so on until the 30th friend.

- **How many cups of water does Rob need to serve?**

40 If 532=151022
924=183652
863=482466
545=202541

- **What does 955 equal?**

41 What is the reason for manholes being round instead of square?

1. 9, 2, 2. To get to the answer, first list down the 3 number combinations that, when multiplied, will give you the product 36. This will give you 8 possible combinations. The riddle says the man went to the house next door to take a look at the house number and came back to tell the woman that he still did not have enough information. This means that 2 of the possible combinations have the same sum, which are:
1 6 6 = 13
9 2 2 = 13
And when the woman says her oldest son plays the piano, you can eliminate the 1 6 6 combination, leaving you with the answer 9, 2, 2.

2. No. Even if you win and the person says they lose and they take your $2, you still lose $1.

3. 6AM. The speed of Earth's rotation will make the difference, not the speed of light.

4. 410. 6+5=111 because 6-5=1, and 6+5=11. 7+3=410 because 7-3=4, and 7+3=10.

5. Since the train Elvis and Marshall are on is moving, the computation is not correct. If it took 5 minutes to meet a second train, but it took the second train an additional 5 minutes to arrive at the point where Elvis and Marshall met the first train, it means the time between trains is 10 minutes. 60 minutes divided by 10 minutes is 6–6 trains arrive in the city each hour.

6. 40 mph
7. 10,100 feet
8. Turn over both hourglasses at the same time. Once time on the 4-minute hourglass runs out, turn it over. Do the same when time on the 7-minute hourglass runs out. When the 4-minute hourglass runs out of time again, it means the 7-minute hourglass has been running for 1 minute. Turn the 7-minute hourglass over and once time runs out on that, 9 minutes has passed.
9. 1, 3, 9, and 27 pounds
10. (7-5)2+96+8-4-3-1=100
11. Mary lost a dollar. She will have to pay the pawnshop the $1.50 before they give her back the $2.00.
12. 180 miles
13. 4 combinations
14. The solution would depend on the outcome of the weigh-in. Number the balls 1 to 12. Place balls 1 to 4 in one of the balance scale's pan, and balls 5 to 8 in the other pan. If they balance, the different ball is in group 9 to 12.

Different Ball is in Group 9 to 12:

Put balls 1 and 2 in one pan and balls 9 and 10 in the other pan. If they balance, the different ball is ball 11 or 12. Weigh balls 1 and 11 against each other. If they balance, the answer is 12. If they do not balance, ball 11 is the different ball. Should balls 1 and 2 vs balls 9 and 10 not balance, then the different ball is either 9 or 10. Weigh ball 1 against ball 9. If they balance, the different ball is 10, if they don't balance, the different ball is 9.

If balls 1 to 4 and 5 to 8 do not balance:

It is important to keep track of which side is heavy for each of the weigh-ins. If 5 to 8 is the heavy side, weigh balls 1, 5, and 6 vs balls 2, 7, and 8. If they balance, the different balls are ball 3 or ball 4. Weigh ball 4 against ball 9, and if they balance, then the different ball is ball 3. If they don't balance, the different ball is ball 4. If balls 1, 5, and 6 vs balls 2, 7, and 8 do not balance, and 2, 7, and 8 weigh heavier, then ball 7 or 8 is the heavy ball, or 1 is a light ball. Weigh ball 7 vs ball 8, and whichever side is heavier is the different ball. If 7 and 8 balance, then 1 is the different ball.

If balls 1, 5, and 6 are heavier, then it means ball 5 or ball 6 is the heavy ball or 2 is the light ball. Weigh ball 5 vs ball 6. The heavier one is the different ball. If they balance, ball 2 is the light ball.

15. Box 1=$1
    Box 2=$2
    Box 3=$4
    Box 4=$8
    Box 5=$16
    Box 6=$32
    Box 7=$64
    Box 8=$128
    Box 9=$256
    Box 10=$489
16. 47
17. 35/70+148/296=1
18. With the mother. The baby is being conceived. If M is Michelle's age, and D is the daughter's age:
    M=D+21 (Michelle is 21 years older) or D=M-21
    D+6=(M+6)/5 In 6 years, Michelle will be 5x older than her daughter.
    (M-21)+6=(m+6)/5
    M=20.25
    D=20.25-21=-0.75 or -9 months
19. Sitti has 3 chickens, 2 cows, and 1 pig.
20. Use factorials
    (0!+0!+0!)!
    =(1+1+1)!
    =3!
    =6
21. Adam puts the guitar pick in a box and puts a lock on it.

When David receives it, he puts one of his own locks on the box and sends it back to Adam. Adam then unlocks his lock and sends the box back to David with his lock still on it. When David receives it, he can then open the lock since he also has the key.

22. Tamia and Jody played the fourth game, and Jody won.
23. If you find the length of their routes, you can see that Chris's route is 6.71 feet and Sean's is 6.40 feet, so Sean will reach the bell before his brother will.
24. Bruno should go first.
25. Four pluses and minuses: 98-76+54+3+21
26. She proposes getting 98 pieces of gold and giving 1 piece each to the third and fifth siblings. Since there are 5 siblings, it is in the interest of the third and fifth siblings to accept the 1 piece of gold instead of getting nothing at all.
27. Taylor, Anna, Nikki, Hayley, then Leighton.
28. 4 cats sitting in a square
29. 3,612
30. 0%
31. Dharma should not take the bet to make sure she has a 100% of taking home $300.00.
32. Manny shook 4 hands.
33. Ty has 31 cards, and Anderson has 9 cards. They trade 1 card.
34. There are days that are double counted like exercise, eating, and sleeping, which all also happen during summer vacation and on the weekends. There are also weekends during summer vacation, so the hours there are being counted twice. Also, school isn't held for 24 hours, so the 4 days for school is more.
35. The train station where Rob got on was 6 stops away from the end of the track. Once the train reached the end of the line, it started going back.

When it went back 1 stop, that was the 7th stop but only 5 stops away from where he got on.

36. The boys sold 12 dozen apples to the first customer for $1.00 each dozen. Bob sold 1 dozen and had 3 apples left. Mark sold 4 dozen apples and had 2 apples left. Bruno sold 7 dozen and had 1 apple left. A second customer came and bought the rest of the apples remaining at $3.00 each. At the end of the day, each boy had $10.00.

37. The coin needs to be flipped twice. Dave and Hanson will need to call either heads or tails, and then on the toss, they will switch (e.g. if Dave picks heads on the first toss, then he will switch to tails on the next toss). They have to keep repeating until one of them wins both tosses to determine who is the winner.

38. John plants 3 bushes in a triangle then the last bush in the middle to form a tetrahedron.

39. 1 cup should be enough since each friend is asking for just half the amount each time.

40. 454585

41. The cover will not slip into the hole if the hole is smaller than the cover, as opposed to if the hole was square. The hole would need to be 50% of the cover for the cover not to fall into the hole.

FINAL WORDS

Thank you so much for taking the time to read my book!

I hope you enjoyed going through these great math riddles as much as I enjoyed writing them (which was a lot).

But do you want to know the best thing about this book?

That it doesn't have to end!

I mean, you have this book forever, which means you can share these math riddles with your friends and family time and time again.

So what are you waiting for?

Get reading, get sharing, and most importantly, get thinking!

www.ingramcontent.com/pod-product-compliance
Lightning Source LLC
Chambersburg PA
CBHW062142280426
43673CB00072B/125